Safe Drinking Water for the Immunocompromised

Immunocompromised persons are more vulnerable than the general population to contaminants in drinking water and include persons with cancer undergoing chemotherapy, cancer survivors, transplant recipients, individuals with HIV/ AIDS or other immune disorders, some elderly, infants, and pregnant mothers. *Safe Drinking Water for the Immunocompromised* provides information about safe drinking water choices for the immunocompromised community, pregnant mothers, and the medical professionals from whom they seek advice. The book serves as a primer on drinking water regulations, water chemistry, as well as the terminology used to describe water contaminants, and includes a glossary that explains the terms and concepts in a straightforward manner.

- Offers a science-based rationale for the acceptable level of a contaminant and sources of drinking water that meet or exceed these levels.
- Presents and explains numerous contaminants of concern found in water.
- Explains the different types of purification for bottled water, including reverse osmosis, distillation, and deionization.

Anthony M. Wachinski, Ph.D., is an author of seven technical books and two volumes of his memoirs. He's a subject matter expert with 52 years' experience on most water-related issues and an Air Force veteran. He served as an associate professor of civil engineering and deputy department head at the USAF Academy. He now commutes between Panama City Beach, Florida, and Baybay City, Leyte, Philippines, to spend time with his wife, Joy.

Safe Drinking Water for the Immunocompromised

Anthony M. Wachinski, Ph.D

CRC Press
Taylor & Francis Group
Boca Raton London New York

CRC Press is an imprint of the
Taylor & Francis Group, an **Informa** business

Designed cover image: Shutterstock

First edition published 2025
by CRC Press
2385 NW Executive Center Drive, Suite 320, Boca Raton FL 33431

and by CRC Press
4 Park Square, Milton Park, Abingdon, Oxon, OX14 4RN

CRC Press is an imprint of Taylor & Francis Group, LLC

© 2025 Anthony M. Wachinski

ISBN: 978-1-032-79776-2 (hbk)
ISBN: 978-1-032-79777-9 (pbk)
ISBN: 978-1-003-49383-9 (ebk)

DOI: 10.1201/9781003493839

Typeset in Times
by Apex CoVantage, LLC

This book is dedicated to Joy, my soulmate, my wife, and my best friend.

Contents

Preface

I'm 77 years old and a five-year prostate cancer survivor – Oorah! I was one of the lucky guys. I was a candidate for CyberKnife, an extremely accurate radiation treatment, no chemo, and minimal radiation impact on my urinary system. My immune system was left intact. I remain immunocompetent. I'm an official "water guy." I've practiced in the water industry for over 50 years. I live in paradise, Panama City Beach, Florida.

After moving here from Long Island, New York, I wanted to make sure that my drinking water was zero-risk for staying cancer-free. So, I did what you should do. I looked up the Consumer Confidence Report (CCR) provided by my municipal water provider. I later chose to have my tap water tested to experience the process in case I ever wrote a book. Let me digress just a little. I was the water commissioner and eventually president of a 100-home development on Long Island. Our water source was the Lloyd Aquifer, one of the purest water sources on the planet. I was responsible for writing and publishing our CCR. Because our water was pristine, and even though we had immuno-compromised (IC) individuals and senior citizens living in the development, I did not include "The Disclaimer," born, I'm convinced, from the *Mission: Impossible* movies, "Mr. Phelps, if you or any of your IM force is caught . . . the secretary will disavow any knowledge of your actions." As water commissioner I answered all questions, in person, including all IC individual's questions about the safety of the drinking water. I did the research for them.

I decided to write this book so that any IC individual can make informed choices about the safety of their drinking water without having to interpret guidance, for the most part, directed to water professionals. I also wrote this book for medical professionals who are asked on many occasions to provide guidance to IC individuals on safe drinking water choices.

By my calculations, there are roughly 36 million IC individuals in the United States. An immunocompromised individual – also referred to as a host in the medical speak – is an individual who does not have the ability to respond normally to an infection because of an impaired or weakened immune system.

I first discovered that immunocompromised (IC) individuals had been abandoned by the water community after my CyberKnife treatment. My family practice doctor recommended I read *Anticancer: A New way of Life*, by David-Schreiber, M.D., Ph.D. I devoured the book and adopted many of his recommendations to stay cancer-free. But something was missing from the book that I didn't catch until five years later. Drinking water quality guidance for cancer patients or individuals undergoing chemotherapy was not mentioned in the book. Why? Later I did an extensive search to find the latest guidance on drinking water for IC individuals. The driver to write this book is born in the statement at the end of a Consumer Confidence Report (CCR) and at the end of a bottled water analysis report:

> Some people may be more vulnerable to contaminants in drinking water than the general population. Immuno-compromised persons such as persons with cancer undergoing chemotherapy, persons who have undergone organ transplants, people with HIV/ AIDS or other immune system disorders, some elderly, and infants can be particularly at- risk from infections. These people should seek advice about drinking water from their health care providers. EPA/CDC guidelines on appropriate means to lessen the risk of infection by Cryptosporidium and other microbiological contaminants are available from the Safe Drinking Water Hotline (800–426–4791).

> The Hotline refers readers to EPA/s safe drinking water act (SDWA).

> I just recently found this same disclaimer at the end of a bottled water analysis report. Interpretation

The elephant in the room is "who sets contaminant levels for IC drinking water?" The EPA sets maximum contaminant levels in drinking water for immunocompetent individuals. It does not recommend any contaminant safe level, nor does the Center for Disease Control (CDC). Bottled water companies post the same statement. No private or government organization has stepped up to recommend safe drinking water options for the immunocompromised.

This book is concerned only with drinking water for immunocompromised individuals. It does not address water quality issues in health care facilities.

Acronym List

ANSI	American National Standards Institute
AWWA	American Water Works Association
CCR	Consumer Confidence List
CDC	Centers for Disease Control and Prevention
Crypto	*Cryptosporidium parvum*
CWS	community water system
DBP	disinfection by-product
E. coli	*Escherichia coli*
EPA	U.S. Environmental Protection Agency
FDA	U.S. Food and Drug Administration
HAA5	five haloacetic acids
GAC	granular activated carbon
IARC	International Agency for Research on Cancer
MCL	maximum contaminant level
MCLG	maximum contaminant level goal
MF	microfiltration
mgd	million gallons per day
mg/L	milligrams per liter
mrem	1/1000th of a rem
MRL	minimum reporting limit
NA	not applicable
ND	not detected at or above the MRL
ND	non detectable
NF	nanofiltration
NOM	natural organic matter
NSF	National Sanitation Foundation
NTU	Nephelometric Turbidity Units
PDWS	Primary Drinking Water Standards
PFAS	per-and polyfluoroalkyl substances
PFOA	perfluorooctanoic acid
PFOS	perfluorooctanesulfonic acid
PHG	public health goal
POE	point of entry
POU	point of use

ppb	parts per billion
ppm	parts per million
ppt	parts per trillion
PWS	public water system
REM	roentgen equivalent man
RO	reverse osmosis
SDWA	Safe Drinking Water Act
SOQ	standard of quality
TDS	total dissolved solids
THMs	trihalomethanes
TOC	total organic carbon
UF	ultrafiltration
UV	ultraviolet
USEPA	United States Environmental Protection Agency
USP	United States Pharmacopeia

Safe Drinking Water Overview

1

OVERVIEW

If you're an IC individual, the question is not "Is my water safe to drink?" but "Is my drinking water safe for me to drink?" This book is written to the 36 million IC individuals living in the United States. It is written for the individual holding down two jobs to support his newly pregnant wife, who doesn't have time figure out what is safe drinking water for her. It is written to the AIDS survivor who needs zero-risk drinking water. It is written to the business executive who just had open-heart surgery, and the neurosurgeon battling cancer and chemotherapy. It is written to every IC individual looking for answers, not guidelines, and not hearsay about safe drinking water. Lastly, this book is written for medical professionals that are asked by their patients for guidance in making safe drinking water choices.

INTRODUCTION

The purpose of this book is to provide to you the IC water customer, the information you need to make informed decisions about your drinking water. IC individuals are more vulnerable than immunocompetent individuals to common waterborne contaminants because they may also be susceptible to disease causing microorganisms that do not usually infect immunocompetent individuals. In medical speak, the reason is a low pathogen burden.

DOI: 10.1201/9781003493839-1

Immunocompromised individuals may experience atypical symptoms or prolonged illness as compared to immunocompetent individuals. They are more susceptible to complications from common infections and more likely to develop bloodstream infection (Gaytán-Martínez et al., 2000) which makes them more vulnerable to infection by *Escherichia coli* (E.*coli*). Human immunodeficiency virus is the most know known IC condition.

This book does not address exposure to chemicals in our daily lives. It does not discuss or recommend diet choices. It does not address food preparation. It does not address personal water well maintenance. It does not address exposure to radiation such as imaging and dental x-rays. It focuses solely drinking water safety and close to zero-risk drinking water options.

WHY THIS BOOK?

The immunocompromised have been forgotten by the safe drinking water community. The driver to write this book was born in the statement at the end of each Consumer Confidence Report (CCR) and at the end of each bottled water analysis report:

> Some people may be more vulnerable to contaminants in drinking water than the general population. Immunocompromised persons such as persons with cancer undergoing chemotherapy, persons who have undergone organ transplants, people with HIV/ AIDS or other immune system disorders, some elderly, and infants can be particularly at risk from infections. These people should seek advice about drinking water from their health care providers. EPA/CDC guidelines on appropriate means to lessen the risk of infection by Cryptosporidium and other microbiological contaminants are available from the Safe Drinking Water Hotline (800–426–4791).

What's wrong with the statement?

A pediatrician wrote in a scientific journal article entitled, "Is the Water Safe for my Baby?":

> pediatricians are confronted by questions about drinking water for which their training may not have prepared them adequately.

The elephant in the room is "Who sets contaminant levels for IC drinking water?" The EPA sets maximum contaminant levels in drinking water for healthy individuals (immunocompetent). It does not recommend any contaminant safe level. Nor does the Center for Disease Control (CDC). Bottled water companies post the same statement. No private or government organization has stepped up to recommend safe drinking water options for the immuno-compromised. At this time the IC community, including pregnant women, have used common sense to drink distilled water or purified bottled water.

Health care providers, that is, medical professionals, receive no training in drinking water quality, safety, or treatment. They may research the topic. This could be both good and bad. One might ask, "Why are the most at-risk individuals in the United States referred to medical professionals with minimal to no training in drinking water quality and safety?" One might also ask, "How good is the information they provide?" An MD's job description is to heal, not to provide expert opinion on drinking water quality to IC individuals.

If you reside in the United States, you get your drinking water from one or more of three sources: municipal drinking water (tap water), bottled water, or a private well.

MUNICIPAL DRINKING WATER (TAP WATER)

I use the same terminology and definitions that the EPA uses, to avoid confusion. In the United States, a public water system provides water for human consumption through pipes or other constructed conveyances to at least 15 service connections or serves an average of at least 25 people for at least 60 days a year. A public water system may be publicly or privately owned.

Public drinking water systems regulated by the EPA and delegated states and tribes provide drinking water to 90 percent of the 340 million Americans. There are over 148,000 public water systems in the United States. EPA classifies these water systems according to the number of people they serve, the source of their water, and whether they serve the same customers year-round or on an occasional basis. EPA defines three types of public water systems, community water systems, non-transient non-community water systems, and transient non-community water systems.

1. Community Water System (CWS): A public water system that supplies water to the same population year-round.
2. Non-Transient Non-Community Water System (NTNCWS): A public water system that regularly supplies water to at least 25 of the same people at least six months per year. Some examples are schools, factories, office buildings, and hospitals which have their own water systems.

3. Transient Non-Community Water System (TNCWS): A public water system that provides water in a place such as a gas station or campground where people do not remain for long periods of time.

Drinking water distribution systems connect water treatment plants or water sources (in the absence of treatment) to customers via a network of pipes, storage facilities, valves, and pumps. In addition to providing water for domestic use, distribution systems may supply water for fire protection, agricultural, and commercial uses. Public water systems (PWSs) are responsible for operating and maintaining their distribution systems, which extend from the designated entry point to the distribution system (EPTDS) – typically the source or water treatment plant – up to the service connection, after which the piping is the property owner's responsibility.

Municipal drinking water is highly regulated by the EPA to ensure that it is potable, that is, safe to drink, and meets or exceeds all legal federal and state enforceable standards. It may not be palatable, that is, taste good, to some individuals because of personal taste sensitivities to a particular water's chemistry. Water sources for municipal drinking water are a surface water, that is, lake, river, or reservoir, or groundwater, for example, an underground aquifer.

The EPA sets legal, health-based limits on over 90 contaminants in drinking water. The legal limit for a contaminant is defined as maximum contaminant level (MCL). It is defined as the level that protects the human health of healthy individuals (immunocompetent) and that water utilities can achieve using the best available technology at a reasonable cost. EPA also sets Maximum Contaminant Level Goals (MLCG) which is the level below which there is no known or expected risk to health. Many MCLGs are zero. EPA also rules how and when water systems should test the drinking water they provide. Cities with populations of 10,000 or greater are required to perform more frequent testing than cities with less than 10,000, notably for *Cryptosporidium parvum* (see the Glossary). More frequent testing is not a solution for IC individuals and pregnant mothers. The risk to IC individuals occurs immediately after an incident, such as a pipe break, and prior to a boil water notice being issued.

In this book I provide the MCL, the MCLG, and my recommended level of contaminants for IC individuals. Am I smarter than the EPA? No. But the EPA does not address drinking water levels for IC individuals. This book is written for IC individuals who, by definition, lack a healthy immune system. I may provide levels stricter than the MCL but higher than the MCLG. Or I might recommend the EPA's MCLG as the level for IC individuals. I do not have to account for a water utility's cost to meet a contaminant level. I'm concerned with your budget to meet the water quality you require.

The Safe Drinking Water Act (SDWA), first passed in 1974 and amended in 1986 and 1996, gives individual states the opportunity to set and enforce their own drinking water standards if the standards are at a minimum as stringent as EPA's national standards. As an example, many states set standards for PFAS before the EPA. (epa.gov).

BOTTLED WATER

Bottled water in the U.S. is regulated as a food by the Food and Drug Administration (FDA). Water sources for bottled water are surface water, groundwater, and municipal drinking water. The surface water source may be a mountain stream, which is not usually the case for municipal drinking water. Although the FDA has defined bottled water classes as Artesian Water, Mineral Water, Purified Water, Sparkling Bottled Water, and Spring Water, the only bottled water choice for IC individuals is Purified Water that meets the definition of "purified water" in the US Pharmacopeia, 23rd edition, Jan 1, 1995. Purified water may also be called "demineralized water, deionized water, distilled water, or reverse osmosis water." I may not recommend demineralized or deionized water because neither process removes dissolved organics.

Bottled drinking water has significant advantages when compared to municipal drinking water. The sources of water and the size or flow rate of the bottling plant compared to a city are the two biggest advantages. High-tech treatment processes like distillation, deionization, and reverse osmosis are economically sound processes to purify bottled water but not cost effective, in most cases, for municipal water. Another advantage is the providers' choice of water source for purified water. Bottled water providers have the luxury to choose the water source, usually clean groundwater. Municipal drinking water providers must treat the source that nature provided. The economics of processing water in a bottled water plant allows ingredients, such as fluoride, to be added as long as the chemicals are listed on the bottle.

One disadvantage in choosing bottled water over municipal water for healthy individuals is the cost differential, which can be 4,000 times higher than tap water. For IC individuals, when risk is added to the equation, the cost differences are closer. You make the choice depending upon risk and your financial situation.

PRIVATE WELLS

Environmental Protection Agency's rules that protect municipal drinking water do not apply to the 15 million households whose source of drinking water is a privately owned well. If your water source is a well, it is your responsibility to ensure your drinking water is safe to drink. The probable water source for your well is groundwater. I cover private well testing in depth in Chapter 5.

A word about common sense choices for IC individuals.

1. If you're IC, drink safe water. Do not drink water directly from shallow wells, lakes, rivers, springs, ponds, and streams.
2. You cannot be 100 percent sure if a tap water contains the protozoa, *Cryptosporidium parvum* (consult the Glossary). Avoid drinking tap water, including water and ice from a refrigerator and drinks made at a fountain, which are usually made with tap water.
3. Public water quality and treatment vary throughout the United States. Use common sense. Assume that all drinking water is not safe for you to drink. Why? Because you are IC, and you cannot afford the wrong choice.

INFORMATION SOURCES

This section addresses reliable information sources about safe drinking water for IC individuals. Fact: Journalists are not considered experts. Information about safe drinking water for IC individuals in articles and books written by journalists may or may not be true. Question these articles. Trust all sources allowed in a court of law, primary and secondary sources, and articles found on Google Scholar. Trust all water related articles written in *Water Research*, *Journal of the American Water Works Association*, and *Environmental Health Perspectives*. Trust articles written on the effects of chemicals on IC individuals in the American Cancer Society, American Chemical Society, National Sanitation Foundation, and National Institutes of Health, BAR.

Trust EPA, CDC, and NIH websites. Trust medical websites on medical topics, not drinking water–related topics. Do not trust hearsay, any magazine article, any newspaper article, any television program, or any movie.

PRIMARY SOURCES

Primary sources include theses, dissertations, scholarly journal articles (research based), some government reports, symposia, and conference proceedings.

SECONDARY SOURCES

These sources offer an analysis or restatement of primary sources. They often try to describe or explain primary sources. They tend to be works which summarize, interpret, reorganize, or otherwise provide an added value to a primary source and include textbooks, edited works, books, and articles that interpret or review research works.

CONSUMER CONFIDENCE REPORT

In 1996, Congress amended the Safe Drinking Water Act (SDWA) and added a provision requiring all community water systems deliver to their customers an annual water quality report. The Consumer Confidence Report (CCR) is a document that provides consumers information about the quality of drinking water in an "easy-to-read" format. The CCR summarizes information that your water system already collects to comply with federal and state regulations. It includes information about the source(s) of water used (i.e., rivers, lakes, reservoirs, or aquifers), levels of various chemical and microbiological contaminants, and any regulatory violations that occurred the previous year.

In March of 2023, EPA proposed new revisions to the Consumer Confidence Report Rule to better communicate drinking water safety risks. The revision is scheduled to go into effect in 2025. Expect changes in 2026.

REFERENCE LIST

Gaytán-Martínez, J., Mateos-García, E., Sánchez-Cortés, E., González-Llaven, J., Casanova-Cardiel, L.J., and Fuentes-Allen, J.L. "Microbiological findings in febrile neutropenia." *Archives of Medical Research* 2000;31(4):388–392.
https://www.cdc.gov
https://www.epa.gov

Water Quality Terminology

2

OVERVIEW

To better understand your drinking water choices, I provide in this chapter a very basic primer on the water chemistry terminology you'll need to understand the rest of the book. To make this an easier read, I've included a glossary of terms. For those who want a more in-depth treatment of basic water chemistry, I recommend *American Water Works Association's Water Supply Operations Series (WSO) Water Treatment Grade 1* (2016). For medical professionals looking to better understand the water chemistry and analytical techniques of water contaminants, I also recommend *Water Treatment Grade 1* to get started, followed by the 24th edition of *Standard Methods for the Examination of Water (2023)* for your office bookshelf. The American Water Works Association (awwa.com) is a trusted source of information on all things water, including peer-reviewed technical papers and an extensive water source bookstore.

INTRODUCTION

Drinking water is water plus everything dissolved in it. It is defined by its physical, chemical, and microbiological characteristics. *Standard Methods for the Examination of Water* (1995) describes, in a technical sense, what could be in and how to analyze for any contaminant in any water including industrial and municipal wastewater (you may know this water as sewage) with directions to chemists and water professionals on how to test for them.

DOI: 10.1201/9781003493839-2

See *Standard Methods* (1995) for a more detailed description of drinking water. Drinking water is defined by its physical characteristics, its dissolved metals and gases, its dissolved inorganic compounds, chlorine residual, organic compounds, and microbiology (bacteria, viruses, protozoa, fungi, and algae).

To a individuals with a compromised immune system, drinking water is water plus anything in it that could have a significant negative impact on their health (for pregnant mothers, her health, and her baby's health).

UNITS OF EXPRESSION

As an IC individual, you are concerned with how much of a particular contaminant is dissolved in your drinking water, that is, the amount of a substance in a specified volume of water. The substance might be particulate matter described as turbidity, fluoride, lead, PFAS, radioactivity, or a microbial contaminant, for example, *E.coli.*

In Chapter 1, you were introduced to the Consumer Confidence Report (CCR), an annual report required by your municipal water authority that describes the sources of drinking water and the levels of various contaminants. These levels are reported as either milligrams per liter (mg/L) or parts per million (ppm), micrograms per liter (ug/L) or parts per billion (ppb), or nano grams per liter, or parts per trillion (ppt). How they are expressed is generally called "units of expression" – the amount of a contaminant dissolved in a specified volume of water.

For purposes of this book:

Milligrams per liter (mg/L) = parts per million (ppm)
Micrograms per liter (ug/L) = parts per billion (ppb)
Nanograms per liter (ng/L) = parts per trillion (ppt)

Radioactivity is expressed differently than inorganic and organic contaminants. Radiation is measured by its rate of decay in decays or disintegrations per second. Curies and becquerels are the standard measures. Thirty-seven billion decays per second is called a curie. One decay per second is called a becquerel. The level of radiation in water is measured in picocuries, one trillionth of a curie, per liter of water. MCLs for dissolved radionuclides such as radium, radon, and uranium are expressed as picocuries/liter (pCi/L).

More important than how much radioactivity is dissolved in water is the dose. Absorbed energy radiation damage is measured in doses. The rem measures the dose (amount) of damage to a human from radiation. Rem stands for radiation equivalent in man. One thousandth of a rem is a millirem or mrem, one millionth of a rem is a microrem, one billionth of a rem is a nanorem, and one trillionth of a rem is a picorem.

A rad is the dose or amount of radiation absorbed by a material. RAD stands for radiation absorbed dose.

The number of microbes in a volume if water is expressed in many ways. The typical volume used is 100 milliliters, abbreviated mLs. One hundred mLs is about 3 1/2 ounces. I'll use a coliform bacteria (an indicator organism) test as an example. Some indicator organisms like fecal coliform are expressed as either coliform absent or coliform present. "Present" indicates that at least one bacterium is present in 100 mLs of water. Presence/absence methods are popular because they are simple, less expensive, and quicker than enumeration methods. They also provide less information about the severity of the bacteria problem that may be needed when trying to determine causes and solutions of bacterial contamination. For a drinking water analysis, coliform present or absent is sufficient.

Many laboratories use a test called the membrane filtration method to test for microorganisms. One hundred milliliters of water are filtered through a membrane filter. The bacteria are captured on the membrane. The filter is next placed in a petri dish with agar (growth medium) to grow the bacteria. If bacteria are present, they appear as colonies on the filter paper that can be counted. The results are reported as the number of colonies per 100 milliliters of water, the number of plaque-forming units per 100 milliliters of water, the most probable number, or as scientific notation, that is, 10^5 organisms per milliliter. Sometimes the bacteria concentration is too high to accurately count, or too numerous to count (TNTC) or confluent. TNTC means that the bacteria concentration was so high that it could not be counted (generally higher than 200 colonies per 100 mL). Confluent means that numerous other bacteria grew on the filter, making identification of coliform bacteria impossible. In either case, another sample is submitted to the laboratory for a more accurate determination.

CHEMICAL SYMBOLS

Each chemical element is designated by a chemical symbol. The symbol consists of either one or two letters. The first letter is always capitalized. Where possible, I've integrated chemical symbols in this chapter's text.

ELEMENTS

The periodic table is a tabular array of the chemical elements organized by atomic number, from the element with the lowest atomic number, hydrogen, to the element with the highest atomic number. Elements are the fundamental materials of which all matter is composed, a substance that can't be broken down by non-nuclear reactions.

IONS

An ion is an atom or group of atoms that has an electric charge. Ions with a positive charge are called cations. Ions with a negative charge are called anions. Many normal substances exist in drinking water as ions. Common examples include sodium, potassium, calcium, chloride, and bicarbonate.

CATIONS

Cations are positively charged ions. Cations present in drinking water are calcium (Ca), magnesium (Mg), manganese (Mn), sodium (Na), potassium (K), and iron (Fe). Lead (Pb), copper (Cu), and other heavy metals may be present in ppb concentrations. Cations of concern to IC individuals are cadmium (Cd), chromium 6 (Cr6), copper (Cu), and lead (Pb). See the Glossary and subsequent chapters.

ANIONS

Anions are negatively charged ions. Anions present in drinking water are chloride (Cl), fluoride (F), sulfate (SO_4) and nitrate (NO_3). The anions of concern to IC individuals are F, NO_3, and nitrite (NO_2) (pregnant mothers) and Cl for individuals with high blood pressure.

COMPOUNDS

Compounds are two or more **elements** bonded together. Compounds are visible to the naked eye Water and table salt are compounds.

INORGANIC COMPOUNDS

Inorganic compounds do not contain carbon atoms. Three main categories of inorganic compounds are acids, bases, and salts. Acids and bases are not discussed in this book. Many inorganic compounds are salts, for example, sodium chloride, sodium sulfate, calcium chloride, calcium sulfate, magnesium chloride, and magnesium sulfate, which are present in drinking water, together with sodium carbonate, calcium carbonate, and magnesium carbonate. Inorganic compounds of concern to IC individuals are the salts of arsenic and heavy metals.

MOLECULES

Molecules are two or more atoms bonded together and invisible to the naked eye. Oxygen gas and nitrogen gas are molecules.

ORGANIC COMPOUNDS

All organic compounds contain carbon atoms combined with one or more elements. The total organic carbon (TOC) test measures for carbon in water. Hydrocarbons contain only carbon and hydrogen. Other organic compounds may contain nitrogen, phosphorous, and sulfur. Synthetic organic compounds of concern contain fluorine or chlorine. At the writing of this book, per- and polyfluoroalkyl substances (PFAS) are of great concern to IC individuals. PFAS are complex synthetic organic chemicals that contain fluorine.

IONIZING RADIATION (RADIOACTIVITY)

A simplified definition of radiation is energy transferred from one place to another, for example, sunlight's heat transferred to your body. (Radiation Basics, 2023). Atoms called radionuclides emit radiation. Radionuclides undergo radioactive decay, release energy, and by doing so transform to a different atom. Decay is measured by a Geiger counter. You've probably heard of the radionuclides, cesium 137, strontium 90, and plutonium 240 in the news. Alpha, beta, and gamma are three types of radiation. Alpha and beta radiation are particles. Gamma radiation is a wave like light.

Everyone is exposed to some natural radiation. Radioactivity, ionizing radiation, is present to some degree in all drinking water. Its concentration and composition depend on the radiochemical composition of the geology through which the raw water may have passed that comes from both cosmic rays and terrestrial sources. The average background dose in the United States is between 100 and 200 mrem/yr. A small proportion of this unavoidable background radiation comes from drinking water that contains radionuclides.

DRINKING WATER CHARACTERIZATION

Physical Properties

The physical properties that describe a drinking water are its pH, total dissolved solids (TDS), turbidity, color, and hardness.

pH is expressed in pH units.

Turbidity is expressed in Nephelometric Turbidity Units or NTUs.

Color is expressed in color units.

Odor is expressed as a threshold odor number or TON.

Total dissolved solids are expressed in mg/L.

Hardness is expressed as mg/L as calcium carbonate or mg/L as $CaCO_3$.

Total suspended solids is expressed as mg/L.

A water's pH describes whether it is acidic, that is, pH less than 7; neutral, pH = 7; or alkaline, that is, pH greater than 7. Most drinking water pH is between 6 and 8. A low pH example is a soft drink cola, about 2. Alkaline water's pH is greater than 8.

Turbidity measures water clarity, that is, cloudiness in Nephelometric Turbidity Units (NTU). It is used to indicate water quality and filtration effectiveness, for example, whether disease-causing organisms are present. Higher turbidity levels in water are often associated with higher levels of disease-causing microorganisms such as viruses, parasites, and some bacteria. Unless there's iron in the water which can color it brown, drinking water's turbidity is less than 3 NTU and usually around 0.1 NTU. Turbidity is seldom a concern in municipal drinking water and not of concern in bottled water quality. I direct the reader to the Safe Drinking Water Act and *epa.gov.* to learn more about the turbidity of US drinking water.

Bottled water is colorless. Municipal drinking water is colorless unless certain safe-to-drink constituents are present. Color is expressed in units from 0 to 500 on the APHA color scale where distilled water is 0. The National Secondary Drinking Water Regulations (NSDWR) standard is a maximum of 15 units, the same as the World health Organization (WHO).

Hardness is a characteristic of water caused by high concentration of calcium and magnesium. Hard water scales cookware and makes lathering difficult. It also discolors bathtub drains. Hard water is safe for IC individuals to drink but, if treated by the ion exchange process, may increase the sodium concentration and is not recommended for individuals with high blood pressure.

The TDS test describes the total concentration of dissolved inorganic constituents. Simply, TDS is the sum of all dissolved matter in a water. Long Island, New York's Lloyd Aquifer's TDS is about 10 ppm. Some bottled water's TDS are as high as 1200 mg/L.

DISINFECTANTS

Chlorine, as hypochlorite, and chloramines are the most common chemicals used to disinfect drinking water.

CHLORINE

Chlorine is a strong oxidizing disinfectant used to treat drinking water.

Chloramines. The EPA defines chloramines as disinfectants used to treat drinking water. Chloramines are most commonly formed when ammonia is added to chlorine. Chloramines provide longer-lasting disinfection as the water moves through pipes to consumers. This type of disinfection is known as secondary disinfection.

DISINFECTION BY-PRODUCTS

All commonly used chemical disinfectants (e.g. chlorine, chlorine dioxide, chloramines, and ozone) react with organic matter and/or bromide to varying degrees to form different disinfection by-products (DBPs). Disinfection by-products (DBPs) are formed during the disinfection process of surface water when the decaying vegetation combines with bromide to form a precursor compound that, when oxidized by the disinfection process, forms disinfection by-products. Decaying vegetation is called natural organic matter. It is expressed as mg/L of total organic carbon(TOC).

Halo acetic acids are disinfection by-products formed during the chlorination of water containing natural organic matter. Acute effects of THMs in drinking water on immunocompetent individuals are rare. Two individual HAAs, chloroform and bromodichloromethane, are suspected carcinogens to humans, but with inadequate evidence.

Trihalomethanes (THMs) are one of the most common disinfection by-products. The principal THMs of concern are chloroform, bromodichloromethane, chlorodibromomethane, and bromoform. These THMs are regulated as a group, with a maximum contaminant level (MCL) established for total THMs (TTHM).

REFERENCE LIST

American Public Health Association, American Water Works Association, and Water Environment Federation. *Standard Methods for the Examination of Water and Wastewater.* 20th ed. Washington, DC: American Public Health Association, 1995.

American Public Health Association, American Water Works Association, Water Environment Federation, Lipps, W.C., Braun-Howland, E.B., and Baxter, T.E., eds. *Standard Methods for the Examination of Water and Wastewater.* 24th ed. Washington, DC: APHA Press, 2023.

AWWA & ABC, WSO. "Basic microbiology and chemistry." In *Water Treatment, Grade 1,* Chapter 1, pp. 1–32, American Water Works Association, Association of Boards of Certification, Denver, 2016.

National Research Council Division on Earth and Life Studies; Commission on Life Sciences; Safe Drinking Water Committee. "Radioactivity in drinking water." In *Drinking Water and Health: Volume 1,* Chapter VII. National Academies Press, Washington, DC, 1977.

Radiation Basics. Presentation to the Nevada Legislature, Carson City, NV, 2023. https://www.leg.state.nv.us

Drinking Water Contaminants

3

OVERVIEW

IC individuals are more vulnerable than immunocompetent individuals to common waterborne contaminants because they may also be susceptible to pathogens that do not usually infect immunocompetent individuals. These pathogens are called opportunistic pathogens.

INTRODUCTION

In this chapter I focus on the contaminants of concern to IC individuals, pregnant mothers, and the elderly. These contaminants are microbial, inorganic, and organic. Microbial contaminants include bacteria, fecal coliform and *E. coli*, the protozoa *Cryptosporidium* and *Giardia*, fungal and viral. *Cryptosporidium* is associated with severe life-threatening illness among immunocompromised individuals. Legionnaires' disease is not usually transmitted through drinking water. Viral infections transmitted through drinking water are extremely rare because of US disinfection practices. Pathogenic fungi are opportunistic and can cause fungal infection disease in patients with immunocompromised conditions.

DOI: 10.1201/9781003493839-3

Microbial Contaminants

Most waterborne pathogens cause gastrointestinal illness with different symptoms in immunocompetent individuals and prolonged illness and death in IC individuals. Microorganisms are a threat to IC individuals. The National Cancer Institute (Cancer 2023) defines a microorganism as an organism that can be seen only through a microscope. Microorganisms are very diverse and include bacteria, fungi, algae, and protozoa. Although viruses are defined as a complex molecule and are not considered living organisms, they are sometimes classified as microorganisms. Parasites are microorganisms that live within, and may cause harm to, other organisms – including humans. Disease-causing microbial contaminants, that is, pathogens, are bacteria, protozoa, and viruses.

Bacteria are single-cell microorganisms between 0.5 and one micron in diameter. To compare, bacteria are smaller than protozoa and larger than a virus. Viruses are submicroscopic infectious agents, a thousand times smaller than bacteria that replicate only inside the living cells of an organism. Viral infections transmitted through drinking water are extremely rare because of US disinfection practices. Protozoa are single-cell organisms larger than bacteria. For example, *Giardia lamblia*'s diameter ranges between 8 and 18 ums. *Cryptosporidium parvum*'s diameter ranges from 4 to 6 um. Protozoan parasites of concern to IC individuals in drinking water are *Giardia lamblia* and *Cryptosporidium parvum*. *Cryptosporidium* is associated with severe life-threatening illness among immunocompromised individuals. Legionnaires' disease is not usually transmitted through drinking water. Most fungal infections are common in humans. Pathogenic fungi are opportunistic but can cause fungal infection disease in patients with immunocompromised conditions, such as malignancy, chemotherapy, transplantation, acquired immunodeficiency syndrome, and usage of immunosuppressant drugs. Most invasive infections are caused by *Aspergillus* species, mucoromycetes, *Cryptococcus* species, and *Candida* species.

Opportunistic bacteria and pathogens are typically non-pathogenic microorganisms that act as a pathogen in certain circumstances. Opportunistic bacteria and pathogens lay dormant for long periods of time until the host's immune system is suppressed (immunocompromised individuals) and cause infection to the host. *Pseudomonas, Aeromonas hydrophila, Edwardsiella tarda, Flavobacterium, Klebsiella, Enterobacter, Serratia, Proteus, Providencia, Citrobacter,* and *Acinetobacter* are prevalent in the environment.

The WSO (2016) states that the pathogenic bacteria of interest to immunocompetent individuals (persons in good health) in drinking water are *Salmonella*, pathogenic *E. coli, Shigella, Legionella, Campylobacter. Shigella* causes bacillary dysentery that is usually not life-threatening but can be in IC. Campylobacter infections result in diarrhea and vomiting.

PATHOGENIC MICROORGANISMS OF INTEREST TO IC INDIVIDUALS

Pathogenic *Escherichia coli* O157:H7 can cause kidney failure and death in immunocompromised individuals. *Escherichia coli* is a pathogen in drinking water linked to a variety of diseases ranging from urinary tract infections, sepsis, meningitis, and bacteremia to diarrhea (Invik, 2017). Bloodstream infection is an important cause of death in immunocompromised patients. In recent years, *E. coli* has gradually become the most common pathogen of bloodstream infection and received extensive attention because of its severe antibiotic resistance.

Cryptosporidium infection is one of the important causes of diarrheal illnesses worldwide. Cryptosporidiosis in humans is usually caused by *Cryptosporidium parvum*. Cryptosporidiosis affects immunocompetent, particularly children under the age of 5 years, and immunocompromised individuals worldwide, especially HIV-infected individuals. It can cause diarrhea lasting about one to two weeks, extending up to 2.5 months among the immunocompetent and a more severe life-threatening illness among immunocompromised individuals (Hunter, 2002). Both parasites *Crypto* and *Giardia* reproduce in the intestine of a susceptible host (humans or animals) and shed environmentally resistant cysts (*Giardia*) or oocyst (*Cryptosporidium*) in their feces. The cysts and oocysts can survive for long periods in the environment (WSO).

Hepatitis A is a vaccine-preventable liver infection caused by the hepatitis A virus (HAV). Private well contamination is possible. Well water can be contaminated with HAV if the well is contaminated with sewage. The well can become contaminated through the fecal matter of infected humans entering a non-sealed well through septic tank effluent and contaminated storm water runoff.

Giardiasis

Giardiasis contamination of well water is rare. Werner Espelage et al. (2010) reported that contact with fresh water or drinking of tap water were not associated with disease in Germany.

Legionnaires' Disease

Legionnaires' disease is a serious type of pneumonia caused by the bacteria *Legionella*. *Legionella* causes pneumonia-like symptoms and causes infection of susceptible hosts which occurs through inhalation from aerosols. It is often found in cooling towers and colonizes plumbing systems. *Legionella* is not typically found in well water.

Protozoa

The protozoan parasites of concern in drinking water are *Cryptosporidium parvum* and *Giardia lamblia*. Common microbial contaminants include fecal coliform and *E. coli*, parasites – *Cryptosporidium* and *Giardia*. *Cryptosporidium* is associated with severe life-threatening illness among immunocompromised individuals. Legionnaires' disease is not usually transmitted through drinking water. Most fungal infections are common in humans. Pathogenic fungi are opportunistic but can cause fungal infection disease in patients with immunocompromised conditions, such as malignancy, chemotherapy, transplantation, acquired immunodeficiency syndrome, and usage of immunosuppressant drugs. Most invasive infections are caused by *Aspergillus* species, mucormycetes, *Cryptococcus* species, and *Candida* species.

Inorganic Contaminants

Inorganic compounds of concern to IC individuals are arsenic (also included as a heavy metal), disinfectants, disinfection by-products, heavy metals, fluoride, nitrates, and radioactive compounds.

Disinfectants

Chlorine

Chlorine is a strong oxidizing disinfectant that is used to treat drinking water. Chlorine itself is not a threat to IC individuals. When the precursors, natural organic matter and bromide, are present in water, all chemical disinfectants react with the natural organic matter and bromide to form disinfection by-products which may prove harmful to IC individuals.

Chloramines

The EPA defines chloramines as disinfectants used to treat drinking water. Chloramines are most commonly formed when ammonia is added to chlorine to treat drinking water. Chloramines provide longer-lasting disinfection as the water moves through pipes to consumers. This type of disinfection is known as secondary disinfection. Chloramines are not a threat to IC individuals.

Chlorine Dioxide

Chlorine dioxide is a disinfectant that kills bacteria, viruses, and *Giardia*. It is effective against *Cryptosporidium*. It also improves taste and odor, destroys sulfides, cyanides, and phenols, controls algae, and neutralizes iron and manganese ions. Chlorine dioxide is not a threat to IC individuals.

Disinfection By-Products

Any disinfection by-products formed during the disinfection process may pose a threat to IC individuals.

Disinfection by-products (DBPs)are formed by the reaction of chemical disinfectants with the by-product precursors, natural organic matter, and bromide. Natural organic matter (usually measured as total organic carbon (TOC)) and inorganic matter (bromide) are the most significant disinfection by-product precursors. All commonly used chemical disinfectants (e.g. chlorine, chlorine dioxide, chloramines, and ozone) react with organic matter and/or bromide to form different disinfection by-products. Trihalomethanes (THMs) are one of the more common disinfection by-products. Disinfection by-products have been deemed safe for immunocompetent individuals. The impact on IC individuals has not been determined because studies to date have been on immunocompetent individuals. Disinfection by-products have been labeled by the International Agency for Research on Cancer (IARC) as Group 2B, that is, "Possibly Carcinogenic to humans." This category is used where there is inadequate evidence of carcinogenicity in humans and sufficient evidence of carcinogenicity in experimental animals. MCL: 80 ppb.

Arsenic

Arsenic in drinking water is carcinogenic to immunocompetent individuals, as classified by IARC and the National Cancer Institute. Dec 7, 2022. MCL: 10 ppb. Public health goal: zero.

Lead

The action level for lead is 0.015 mg/L (15 ppb). The public health goal is zero. Infants who drink formula prepared with lead-contaminated tap water may be at a higher risk of exposure because of the large volume of water they consume relative to their body size. Even though your tap water is lead-free from the water authority, there may be lead in your service line or household plumbing.

You cannot see, taste, or smell lead in drinking water. The only way to know the level of lead in your drinking water is to analyze the water coming out of the tap that you use to reconstitute baby formula. Again. Until you know, I recommend purified bottled water with a lead level 5 ppb or ND.

The CDC recommends the following: Your local water authority and the published CCR is always your first source for testing and identifying lead contamination in your tap water. Ask your water provider if you have a lead service line providing water to your home. If you have a lead service line, ask if there are any programs to assist with removal of the lead service line going to your home. Understand that any work, such as water main or service line replacement, could increase exposure to lead while the work is ongoing and for up to six months after the work is completed.

Ask to have your water tested. Many public water systems will test drinking water for residents upon request. There are also laboratories that are certified to test for lead in water. Understand that water sampling results can vary depending on the time of day, season, method of sampling, flow of water, and other factors.

Lead in drinking water has negative effects on pregnant mothers, fetus, and small children and causes kidney problems in adults. Lead is associated with birth defects, delivery complications, fetal neurotoxicity, and skeletal abnormalities. MCL: 15 ppb. Public health goal: zero.

Nitrates

Risk of nitrates above MCL is to infants below age 6. Worst consequence is blue baby syndrome.

MCL and public health goal is 10 mg/L.

Heavy Metals

The association between heavy metal toxicity and the immunocompromised has not been studied in any detail. The following MCLs and public health goals are for immunocompetent individuals:

Antimony MCL: 6 ppb, PHG: 6 ppb
Arsenic MCL: 10 ppb, PHG: zero
Cadmium MCL: 5 ppb, PHG: 5 ppb
Chromium MCL: 100 ppb, PHG 100 ppb
Copper MCL: 1.3 mg/L, PHG: 1.3 mg/L
Mercury MCL: 2 ppb, PHG 2 ppb
Selenium MCL: 50 ppb, PHG 50 ppb

Fluoride

No guidelines exist that address infant formulae for fluoride consumption by infants. Both milk-based and soy-based formulae contain small amounts of fluoride. In an article published in the *Australian Dental Journal*, researchers determined that the highest concentration of fluoride in water used to reconstitute baby formula with no chance of dental fluorosis was 0.1 mg/L. MCL: 4.0 mg/L. PHG: 4 mg/L.

Organic Contaminants

Organic compounds of concern to IC individuals are PFAS, pesticides, and volatile organics.

PFAS

At the writing of this book, the risk of PFAS, PFOA, and PFOS to immune competent individuals was uncertain. The State of California classified PFOS as a carcinogen in 2021. On March 14, 2023, EPA's science advisory board determined that both PFOA and PFOS are likely to cause kidney and liver cancers. EPA is proposing a National Primary Drinking Water Regulation (NPDWR) to establish maximum contaminant levels (MCLs) for six PFAS in drinking water. PFOA and PFOS as individual contaminants, and PFHxS, PFNA, PFBS, and HFPO-DA (commonly referred to as GenX chemicals) as a PFAS mixture. EPA is also proposing health-based, non-enforceable maximum contaminant level goals (MCLGs) for these six PFAS. PFOA 4.0 ppt (ng/L) and PFOS 4.0 ppt. Some experts say the cutoff for total PFAS levels should be even lower, 1 ppt.

In March 2021, EPA issued a final regulatory determination to regulate perfluorooctanoic acid (PFOA) and perfluorooctanesulfonic acid (PFOS) as contaminants under Safe Drinking Water Act (SDWA). EPA is proposing for public comment a drinking water regulation that includes six PFAS. EPA is proposing to establish MCLGs and an NPDWR for these PFAS in public drinking water supplies. EPA proposes MCLGs for PFOA and PFOS at zero (0) and an enforceable MCL for PFOA and PFOS in drinking water at 4.0 ppt.

Through this action, EPA is also proposing a National Primary Drinking Water Regulation (NPDWR) and health-based maximum contaminant level goals (MCLG) for these four PFAS and their mixtures as well as for PFOA and PFOS. Exposure to these PFAS may cause adverse health effects, and all are likely to occur in drinking water.

The proposed MCLs for PFOA and PFOS are 4 ng/L (individually), and the proposed MCL of an HI of 1.0 with PFAS levels that exceed the proposed MCLs would need to take action to provide safe and reliable drinking water. These systems may install water treatment or consider other options such as using a new uncontaminated source water or connecting to an uncontaminated water system. Activated carbon, anion exchange (AIX), and high-pressure membrane technologies have all been demonstrated to remove PFAS, including PFOA, PFOS, PFHxS, HFPO-DA and its ammonium salt, PFNA, and PFBS, from drinking water systems. These treatment technologies can be installed at a water system's treatment plant and are also available through in-home filter options.

REFERENCE LIST

Espelage, W., et al. "Characteristics and risk factors for symptomatic *Giardia lamblia* infections in Germany." *BMC Public Health* 2010;10, Article number: 41.

Hunter PR, Nichols G. Epidemiological and clinical features of Cryptospordium infection in Immunocompromised patients. *Clin Microbiol Rev.* Chicago, 2002;15:145–154.

Invik, J., Barkema, H.W., Massolo, A., Neumann N.F., and Checkley, S. "Total coliform and Escherichia coli contamination in rural well water: analysis for passive surveillance." *Journal of Water and Health*, IWA Publishing, London, 2017;15:729–740.

WSO. *Water Treatment Grade 1.* American Water Works Association, Association of Boards of Certification, Denver, 2016.

Bottled Water

4

OVERVIEW

Chapter 4 is the CliffsNotes version of bottled water choices for the immunocompromised. It's a roadmap explaining how to choose the bottled water best for your situation. From a water quality point of view, bottled water purified by reverse osmosis, distillation, and in some cases deionization meets or exceeds mine and others' recommended safe-for-the-immunocompromised water quality levels.

INTRODUCTION

In this chapter I provide bottled water choices for the IC community. I do not provide a comparative analysis of tap water versus bottled water. I do not discuss energy requirements to produce bottled water. I do not discuss greenhouse emissions. I do not opine on the ongoing debate about which water is safer to drink, tap water or bottled water. I do not opine on the bottled water industry's impact on the environment. I do not opine on plastics recycling. I do address IC cost considerations.

REGULATORY

A simplified summary of the regulatory landscape for bottled water is provided. Bottled water in the U.S. is regulated as a food by the Food and Drug Administration (FDA). fda.gov. FDA has established specific regulations for bottled water in Title 21 of the *Code of Federal Regulations* (21 *CFR*). I direct the reader to *Code of Federal Regulations*, Title 21, Volume 2, Part 165, Section 165.110. For those new to federal regulations, Title 21 – Food and Drugs, Chapter I-Food and Drug Administration, Department of Health and Human Services, Subchapter B-Food for Human Consumption.

The Federal Food Drug and Cosmetic Act (FD&C) gave the FDA responsibility to ensure that the quality standards for bottled water are compatible with EPA standards for public drinking water. Standard of quality regulations (21 *CFR* §165.110[b]) establish allowable levels for contaminants (chemical, physical, microbial, and radiological). Each time EPA establishes a standard for a contaminant, FDA either adopts the standard or determines that is not required for bottled water. In many cases bottled water standards are tighter than EPA standards, largely because bottled water plants do not require pipelines to deliver the treated water to its customer.

The FDA established Current Good Manufacturing Practice (CGMP) regulations for the processing and bottling of bottled drinking water (21 *CFR* part 129). I encourage IC readers to google 21 CFR Part 129, "Processing and Bottling of Bottled Water" to read for themselves the regulations. My recommendation for the water quality that minimizes risk to most IC individuals is purified water.

WATER SOURCES

Water sources for bottled water are groundwater, that is, springs, and wells, surface water, and municipal drinking water. The surface water source may be a mountain stream, which is not usually the case for municipal drinking water. My recommendation is in line with others who recommend purified bottled water that meets the definition of "purified water" in the *US Pharmacopeia*, Revision 23rd edition, Jan 1, 1995 as the only water choice for many IC individuals. Purified water may also be called reverse osmosis treated water, distilled water, demineralized water or deionized water. Demineralized or deionized water may not remove dissolved organics. Purified water is prepared from water that meets EPA primary drinking water standards. It is further processed by either reverse osmosis, distillation, deionization, or demineralization.

Surface water is not a bottled water source that I recommend. It is seldom a source because of its water quality. Municipal drinking water is not usually a source that I recommend to IC individuals, especially if the source was surface water, because of its quality and risk of disinfection by-products.

Any bottled water provider that chooses surface water or groundwater under the influence of surface water is regulated by the EPA and must follow EPA regulations on its treatment. Bottled water prepared using surface water or groundwater under the influence of surface water (GWUI) source poses a risk to IC individuals. I recommend that IC not consider any bottled water that uses these sources.

UC Davis Medical Center in 2017 listed the following brands as safe for IC individuals: Alhambra, Aquafina, Crystal Geyser, Dasani, Kirkland, Nestlé, Safeway, and WinCo. Visit IBWA's website, www.bottledwater.org. Call 1–800–928–3711 for additional safe brands.

RISK

The choice of drinking water for an IC individual is risk-based. The greatest risk is microbial contamination. For a municipal drinking water, the greatest risk occurs immediately following a pipe break or contamination event before a boil water notice is issued, which can be up to 48 hours. Bottled water is delivered in sealed containers and eliminates this risk.

I recommend bottled water over municipal drinking water for the IC that can afford it. Bottled drinking water production has a number of advantages over municipal drinking water. These advantages are choice of source water, volume of source water treated, cost of water, and method of delivery. Bottled water companies have the option to choose a higher quality source water, usually a clean groundwater. Municipalities must use and treat the geographical water source. Municipal water customers do not have a source water choice.

High-tech treatment processes like distillation and reverse osmosis are economically sound processes to purify bottled water but may not be cost-effective, in most cases, for municipal water. The cost to treat municipal drinking water has a price ceiling often regulated by state and local governments, which limits treatment technology options. The cost of bottled water is market focused. Customers have a choice. Lastly, municipal drinking water leaving the plant must be chlorinated to satisfy the chlorine demand of the water and maintain a chlorine residual that ensures no microbiological activity at the tap but a risk of forming disinfection by-products. It is then delivered via miles of underground pipes subject to leaks and accidental failures – an unacceptable risk to IC individuals. Bottled water is delivered in sanitary sealed containers that were filled under controlled conditions at the bottling plant.

CHOOSING A BOTTLED WATER

Bottled water choices can appear overwhelming. Bottled water choices for IC individuals are not. A safe estimate is that there are over a hundred bottled water companies operating in the US offering a variety of bottled water choices. An IC individual's only bottled water choice is purified still water. Purified water, that is, bottled water purified by reverse osmosis or distillation, is an acceptable drinking water for all IC individuals.

From a water quality point of view, bottled water is pathogen-free. Surface water and groundwater under the influence of surface water are not approved as bottled water sources. Bottled water purified by reverse osmosis or distillation is *Cryptosporidium*-free and meets or exceeds my safe-for-the-immunocompromised recommended water quality.

The CDC includes one-micron absolute filtered water as purified. I do not consider one-micron absolute filtered water as purified because it does not remove organics, viruses, or PFAS. I do not include demineralized water or deionized water as purified water for the same reason.

The trade association for the bottled water industry is the International Bottled Water Association (IBWA), bottledwater.org. The IBWA says that it supports federal limits for PFAS and that bottled water should have PFAS levels below 5 ppt for any single compound and 10 ppt for more than one. I recommend their website for individuals who are interested in learning more about the bottled water industry.

Commercially bottled water labels reading "well water," "artesian well water," "spring water," or "mineral water" do not guarantee that the water is *Cryptosporidium parvum*–free. However, commercially bottled water that comes from protected wells or protected springs is less likely to contain *Cryptosporidium parvum* than from less protected sources, such as a river, reservoir, or lake. Any bottled water labeled, treated by

- reverse osmosis,
- distillation
- absolute 1 micron or smaller, and
- "one-micron absolute"

(no matter what the source) should have removed or inactivated *Crypto*. If you drink commercially bottled water and are concerned about *Cryptosporidium parvum*, read the label and look for this information.

Bottled water quality reports, also called bottled water analysis reports, are provided by bottled water companies and are available online. Report format and thoroughness vary with brand as does price. My suggestion when choosing a brand is to evaluate by performance and then price. Compare the analysis reports for a number of brands, and select the brands that meet your performance criteria. As an example, fluoride concentration for a bottled water used to prepare baby formula should be less than or equal to 0.1 ppm where fluoride concentration for another IC individual may not be of concern.

ECONOMICS

Choosing to drink bottled water instead of municipal drinking water for immune competent individuals is more expensive than the cost of municipal water. The cost differential can be as much as 4,000 times higher than tap water. But when the risk of infection is added to the equation, the economics are closer. Still, purified bottled water ranges in cost per ounce from 0.9 to 5 cents. Your choice depends upon risk and your finances.

WATER QUALITY

Bottled drinking water production has advantages when compared to municipal drinking water. The two biggest advantages compared to municipal drinking water are choice of water source and bottling plant flow rate. Municipal drinking water providers must treat the geographical source. Bottled water companies have more freedom to locate near a high-quality water source. High-tech treatment processes like distillation and reverse osmosis are economically sound processes to purify typical bottling water plant flows but not cost-effective, in most cases, for municipal water. The economics of processing water in a bottled water plant allows ingredients, such as fluoride, to be added but must be labeled.

CHOICES AND BRANDS

There were over 100 brands of bottled water in the US in 2002.

REFERENCE LIST

Code of Federal Regulations, Title 21, Volume 2, Part 165, Section 165.110.

The United States Pharmacopeia Revision 23. *The National Formulary 18: Official from January 1, 1995*. Rockville, MD: United States Pharmacopeial Convention, ©1994.

Private Groundwater Wells

5

OVERVIEW

Over 15 million U.S. households receive their drinking water from private groundwater wells. Environmental Protection Agency regulations that protect public drinking water systems do not apply to these wells. Typically, private water systems that serve no more than 25 people at least 60 days of the year and have no more than 15 service connections are not regulated by the EPA.

IC individuals who rely on private wells for their drinking water are at high risk of microbial infection. They alone are responsible for ensuring that their well water is safe to drink. All private wells use groundwater. Groundwater can be contaminated by failed septic tanks, leaking underground, heating oil tanks, fertilizers, pesticides, and runoff from urban areas if the well is not properly sealed. An IC individual cannot risk consuming polluted groundwater. If contaminated groundwater is consumed by an IC individual, it could cause severe long-lasting illness.

INTRODUCTION

In this chapter I discuss private well water quality and well treatment scenarios. I do not address well construction, well maintenance, or groundwater protection. I do have a few simple recommendations. I recommend contacting

---

either your local health department or registered well driller to, as a minimum, inspect your well to ensure that your well is properly sealed. A well system that is not properly sealed is vulnerable to microbiological contamination. You may want to extend a buried well casing above the ground and slope the ground away from the casing to prevent surface water from entering the well. Also, make sure the top of the casing has a tight sanitary well cap that prevents insects and surface water from entering.

Go to wellowner.org to find a qualified professional and to receive a copy of the National Ground Waters Association's (NGWA) *Well Owners Guide*. Health or environmental departments or county governments should have a list of the state-certified (licensed) laboratories in your area that test for a variety of substances. See Well Water Information Based on Where You Live (United States Environmental Protection Agency) and State Certified Drinking Water Laboratories (United States Environmental Protection Agency) are EPA links for certified testing laboratories in your state. You can just google using the keywords "private well water testing services near me."

SAMPLING

The fundamental first step in analyzing your drinking water is to collect a representative sample of your well water. Common sense and *Standard Methods* tell us that an analytical test result is no better than the sample on which it was performed. Either you or a laboratory technician is responsible to collect the water sample for your well testing. My strong recommendation is to rely on a trained professional to do the sampling to ensure that a representative sample is collected. These water analyses are not cheap. What you want to avoid is paying hundreds of dollars for a well water analysis only to discover that the sample was not properly taken and the results are in question.

Proper water testing for bacteria requires that you obtain a sterilized sample bottle from the laboratory and collect the sample strictly according to their instructions. Failure to properly collect the sample in a sterile container may cause bacteria to be introduced during the sampling process.

Contaminants of concern to IC folks are often present at very low concentrations, that is, the ug/L (ppb) level further complicating sample collection. Sample collection takes on new meaning when collecting PFAS samples. Cook and O'Reilly (2021) describe in their published article in the American Bar Association that collecting a PFAS sample is a constant challenge because PFAS is everywhere.

TESTING

I assume that your well was tested and that you have access to the water analyses. Contaminants of concern to IC individuals are, number one, microbial total coliforms – a positive coliform test signals the possibility of pathogens, such as *E. coli, Cryptosporidium parvum, Salmonella* (can be present in a contaminated groundwater well after a flood or natural disaster, but is more of a food pathogen than a water pathogen) and Legionnaires' disease. Hepatitis A is a vaccine-preventable liver infection caused by the hepatitis A virus (HAV). Private well contamination is possible. Well water can be contaminated with HAV if the well is contaminated with sewage. The well can become contaminated through the fecal matter of infected humans entering a non- sealed well through septic tank effluent and contaminated storm water runoff. Giardiasis contamination of well water is rare. Werner ESPELEGE reported that contact with fresh water or drinking of tap water were not associated with disease in Germany. *Legionella* is not typically found in well water. *It is most often found in large building cooling towers and plumbing systems.*

Typically, well water is first sampled for total coliform bacteria. A test that is positive for coliform bacteria indicates that the conditions exist in the well to support other types of bacteria, including pathogenic bacteria. Most labs will flag that your well is contaminated with a positive coliform test. The next step is to identify all pathogenic organisms in your well water. The presence of *E. coli* in drinking water indicates fecal contamination. When coliform bacteria, *E. coli*, or both are found in your well water, do not drink the well water. Drink bottled water – see Chapter 4. If a microbial free source is not available, boil your water and cool it in a refrigerator. If you can afford it, retain an individual or company that performs maintenance and corrects any well problems. Use another source of water such as purified bottled water for drinking, cooking, preparing baby formula or food, washing produce, brushing your teeth, or any other use where you may swallow the water. Boil water and cool it in your refrigerator to make ice. Ice can cause infection. Use another water source such as purified bottled water if heavy metals, more specifically lead, cadmium, or mercury, hexavalent chromium, are detected. Use another source if inorganics, for example, arsenic, chromium, copper, lead,, fluoride – not common in most wells – nitrites and nitrates for pregnant mothers, pesticides, radioactivity, volatile organic compounds, PFAS, or any carcinogen is detected. IC individuals should not drink the well water and drink purified bottled water until the well water is safe and free from any of these contaminants.

At the writing of this book, PFAS is emerging as the next contaminant of concern to IC individuals. I recommend location, location, location as a private well PFAS well testing strategy. If you live in Alabama, California, Florida, Maryland, Massachusetts, New Jersey, Ohio, or Washington, or near a military base where AFFF was extensively used or near any highly industrialized area, one option is to pay for the "first round PFAS testing" which costs about $75. If you know that your well is PFAS contaminated but the concentration is unknown, again, pay for the first round test.

TESTING FREQUENCY

Private groundwater well's annual test frequency and subsequent testing after a flooding event is based upon the risk to immunocompetent individuals. If you are an IC individual and rely on a private groundwater well for your drinking water and cooking, I recommend quarterly coliform testing and subsequent testing after a flooding event. If your initial water test indicates that total coliform bacteria are present, do not drink from this well. Additional tests for fecal coliform. *E. coli* bacteria and hepatitis A should be performed.

WELL TREATMENT

Selection of a treatment option for a contaminated well is dependent on the contaminant. Before making any costly decisions about your water supply, make sure the coliform bacteria result you have received is accurate. Make sure you used a certified water-testing laboratory and that you carefully followed the sample collection procedure using a sterile sample bottle. You may want to submit a second sample just to confirm the initial result. Also, if you only had a presence/absence test done, you may want to consider asking the lab to count (enumerate) the bacteria in your water.

Activated carbon, anion exchange (AIX), and high-pressure membrane technologies have all been demonstrated to remove PFAS, including PFOA, PFOS, PFHxS, HFPO-DA and its ammonium salt, PFNA, and PFBS, from drinking water systems. These treatment technologies are also available through in-home filter options (see Section XI of this preamble for additional discussion on available treatment technologies).

REFERENCES

Cook, L., and O'Reilly, K. *Regulating PFAS at the Edge of Detection.* American Bar Association, Chicago, June 2021.

Espelage, W., et al. "Characteristics and risk factors for symptomatic *Giardia lamblia* infections in Germany." *BMC Public Health*, Springer Nature, Berlin, 2010;10, Article number: 4.

Immuno-compromised Individuals

6

OVERVIEW

In this chapter I explain the immunocompromised individual in more detail. I reiterate my concern about the lack of recommendations to IC individuals on drinking water choices. I concentrate on the higher-risk microbial, inorganic, and organic contaminants and discuss commonsense behavior and provide mitigation strategies.

I'm a five-year prostate cancer survivor. I was one of the lucky guys. I was a candidate for CyberKnife, an extremely accurate radiation treatment, no chemo, and minimal radiation impact on my urinary system. After my treatment, my family practice doctor recommended I read *Anticancer: A New way of Life*, by David-Schreiber, M.D., Ph.D. I devoured the book in one sitting. After reading it, I left nothing to chance. I eliminated as many chemical-causing agents as I could. I radically changed my diet, all but eliminating sugar. No more beef. Reflecting as I write this chapter. Something was missing from the book. Drinking water. Maybe because I'm a water guy and subconsciously didn't want any drinking water advice. Maybe because my drinking water source at the time was the Lloyd Aquifer with its exceptional water quality? There was no mention of water in the book. As I researched this chapter, again, nothing substantial on drinking water quality for cancer patients or individuals undergoing chemotherapy.

What's disturbing to me is the number of organizations with minimal, if any, training or education in drinking water chemistry providing online guidance to immunocompromised individuals about safe drinking water.

DOI: 10.1201/9781003493839-6

IMMUNOCOMPROMISED EXPLAINED

I understand that if you're an IC individual, that you need no explanation of what it's like to be immunocompromised. This paragraph is to explain my rationale in drinking water choices and mitigation strategies. An IC individual – also referred to as a host in medical speak – is an individual who does not have the ability to respond normally to an infection because of an impaired or weakened immune system. Microorganisms are a threat to IC individuals. Immunocompromised individuals and patients are susceptible to bacterial, fungal, viral, and parasitic infections that immunocompetent persons usually overcome. They are more vulnerable than to common waterborne contaminants because they may also be susceptible to pathogens that do not usually infect immunocompetent individuals. Most waterborne pathogens cause gastrointestinal illness with different symptoms in immunocompetent individuals but possible prolonged illness and death in IC individuals. Diagnosis may be challenging because IC folks are often infected by opportunistic pathogens that a physician may not consider. IC patients, usually suffering from immunosuppression, chemotherapy, particularly steroids, stem cell transplant, and long-term hospitalization simultaneously, are more likely to develop bloodstream infection (Gaytan-Martinez, 2000; Tao et al., 2020).

FIRST STEPS

Here's my proposed safe drinking water lifestyle change. Getting infected by the protozoan parasite *Cryptosporidium parvum* and/or the bacteria *Escherichia coli* O157:H7 is not an option under any circumstance. Said a different way, not ever getting infected by *Crypto* and/or *E. coli* is your highest priority. Here's my commonsense list.

1. Always drink safe water. Never drink water directly from shallow wells, lakes, rivers, springs, ponds, and streams.
2. Since you cannot be 100 percent sure that tap water is free of *Crypto*, never drink tap water with ice from a refrigerator and drinks made at a fountain, which are usually made with tap water.

3. For immunocompetent individuals, drinking water in the United States is safe to drink in any state. For IC individuals, assume that all drinking water is not safe for you to drink. Why? Because you cannot afford the wrong choice.

MICROBIAL CONTAMINANTS – A QUICK REVIEW

The National Cancer Institute (https://www.cancer.gov/publications/dictionaries/cancer-terms/def/microorganism) defines a microorganism as an organism that can be seen only through a microscope. Microorganisms are very diverse and include bacteria, fungi, algae, and protozoa. Viruses are defined as complex molecules or a strand of RNA or DNA and are not considered living organisms, but they are sometimes classified as microorganisms. Parasites are microorganisms that live in a host organism and get its food from the host and may cause harm to other organisms – including humans. Disease-causing microbial contaminants, that is, pathogens, are bacteria, protozoa, and viruses.

The drinking water protozoan parasite of most concern to IC individuals is *Cryptosporidium parvum*. The bacteria contaminant of most concern is *Escherichia coli* O157:H7.

Cryptosporidiosis, associated with severe life-threatening illness among immunocompromised individuals, is caused by the parasite *Cryptosporidium parvum*, also known as Crypto. Cryptosporidiosis affects immunocompetent and immunocompromised individuals, especially HIV-infected individuals. It can cause diarrhea lasting from one week to over 2.5 months among the immunocompetent and a more severe life-threatening illness among immunocompromised individuals (Hunter & Nichols, 2002).

Escherichia coli, associated with severe antibiotic resistance and death in immunocompromised patients, is a pathogen in drinking water linked to diseases ranging from diarrhea to bloodstream infection (Invek, 2017). Pathogenic *Escherichia coli* O157:H7 can cause kidney failure and death in immunocompromised individuals.

Legionnaires' disease is a serious type of pneumonia caused by the bacteria *Legionella*. Legionnaires' disease is not usually transmitted through drinking water.

Pathogenic fungi are opportunistic but can cause fungal infection disease in immunocompromised patients.

MICROBIAL MITIGATION STRATEGIES

Safe drinking water is available to every IC individual regardless of budget. *Crypto* and *E. coli* are killed when drinking water is boiled. The practice of boiling tap water before drinking also guarantees against an IC individual ever drinking contaminated water before a boil water notice is posted. Boiling water may remove a dissolved volatile organic contaminant. It will not remove PFAS or any inorganic contaminants such as lead. Pregnant mothers, see Chapter 7.

Here's my recommended boil water strategy.

Carefully, repeat, carefully boil all tap water for one minute of vigorous boiling and store in a clean dishwater, heat-resistant plastic, or other suitable container and cool in a refrigerator or at room temperature. The boiled water is now safe for drinking, brushing teeth, and making ice. There is debate about the length of time water should be boiled. I Abubaker (2007) recommends boiling all water, from whatever source, that might be consumed by immuno-compromised persons for one minute and allowed to cool before use. I agree. My advice is that all IC individuals and patients with HIV infection boil water for a one-minute vigorous boil.

The best other mechanical option is a POU filter rated as **1 micron absolute** or smaller, tested and certified to NSF/ANSI Standard 53 or NSF/ANSI Standard 58 for cyst removal. If you choose this option, you must replace the filter as recommended by the vendor and exercise caution to not infect yourself during the process.

Another option if your budget allows is to use purified bottled water for drinking, making ice, and brushing your teeth. Many folks choose to boil the water used to make ice and purified bottled water for drinking and brushing teeth. Purified bottled water may be your only choice if your tap water is contaminated with inorganic or organic contaminants.

Ice can cause infection. Confirm if the filter on your refrigerator's ice maker is a I micron absolute ANSI/NSF certified to standard 53 or 58. Until you know for sure, use ice trays and your cooled boiled water. Consult your refrigerator manufacturer and Amazon for NSF certified options. Do not neglect filtering your refrigerator water.

Only brush your teeth with cooled boiled water, absolute filtered water, or purified bottled water. POU nanofiltration or reverse osmosis is an option. Consider its purpose, and choose what's best to ensure zero risk for you.

Another option is to combine boiled/cooled water with purified bottled water. Drink purified bottled water. Use boiled/cooled water for everything else.

INORGANIC CONTAMINANTS OF CONCERN TO IC INDIVIDUALS

Inorganic compounds of concern to IC individuals are arsenic (also included as a heavy metal), disinfectants, disinfection by-products, heavy metals, fluoride, nitrates, and radioactive compounds.

My interpretation of the medical literature is that IC individuals have an increased risk of cancer. Please confirm with your MD or do your own research. My advice is ND on all inorganic carcinogens.

The lowest risk option for drinking water is purified bottled water. Consult your CCR to determine the concentration of any inorganic compounds of concern in your drinking water. Don't forget that the CCR's data is for water leaving the treatment plant, not your tap water. Tap water testing is much mor common and affordable. Check local laboratories and online. If you use an online service, read and follow their directions carefully. I'd also google the best tap water sample collection. Consider having your tap water analyzed for the following: aluminum, arsenic, barium, cadmium, chromium, copper, lead, mercury, nickel, strontium, uranium, nitrate, and fluoride if you're a pregnant mother, total organic carbon, and disinfection by-products. All should be ND except Fluoride. Beware of testing companies that try to sell you a water softener. Make sure that your drinking water is safe. A softener may be appropriate if your heavy metals, especially lead, are not N. If your inorganics are ND, deal with softening later. Check with your neighbors to see if they are on a softener.

Fluoridation of water has received great scrutiny but appears to pose little or no cancer risk.

If the source of your municipal drinking water is a surface water, that is, lake, river, or reservoir, determine the TOC. If the TOC is 2 mg/L or greater, assume that you may be at risk from the DBPs formed and consider NSF/ANSI NF/RO and purified bottled water.

Arsenic is carcinogenic to immunocompetent individuals as classified by IARC and the National Cancer Institute. Assuming that IC individuals have an increased risk, arsenic in your drinking water should be ND.

The public health goal for lead is zero. Even though tap water is lead-free when it leaves the water authority's treatment plant, there may be lead in your service line or household plumbing. You cannot see, taste, or smell lead in drinking water. The only way to know the level of lead in your drinking water is to analyze the water coming out of the tap that you use for drinking. Until you know, I recommend purified bottled water with a lead level 5 ppb or ND.

Treatment options are NSF/ANSI 53 lead certified filters, nanofiltration, and reverse osmosis and purified bottled water. The best option if you can afford it is almost always purified bottled water.

The effects of heavy metals on IC individuals have not been fully studied. Cadmium and mercury are known carcinogens and should be ND in your drinking water. Treatment options are NSF/ANSI POU nanofiltration and reverse osmosis, POU ion exchange to remove all heavy metals, and POE softening which would be effective in removing heavy metals and lead.

The presence of pesticide residues in drinking water can be a major risk to maternal and fetal well-being. Pesticides are ND in most municipal drinking water. Pesticides are ND in purified bottled water. If your water comes from a private well, pesticides may be present. Your well water should be tested.

Radioactivity, ionizing radiation, is present to some degree in all drinking water. The dose commitment from radioisotopes in U.S. drinking water supplies is very low. In a hypothetical study (NRC, 1977), a total-body dose of less than one-thousandth of a rem (0.244 mrem) per year would be accumulated. This is less than 1 percent of the 100 to 200 mrem background.

Radiation exposure is a risk in some states, but the greater risk to IC patients is obviously from the radiation treatment. Check your CCR for radioactivity. Your CCR indicates the presence of uranium or radon in your drinking water. Take your CCR's radioactivity level with you and show it to the technician so she/he is aware of your exposure.

MITIGATION STRATEGY

The greatest threat to IC individuals from their drinking water is microbial. *Crypto* and *E. coli* are the greatest threat. The mitigation strategy I've presented for zero microbial risk should approach zero risk for opportunistic pathogens. Boiling your drinking water has little to no effect on inorganic contaminants. Microbial pathogens have an immediate and life-threatening effect on IC individuals. Inorganic contaminant effects are longer term. Here are options.

Have your drinking water analyzed. If you're higher than ND on any carcinogen or suspected carcinogen, use purified bottled water for drinking and brushing your teeth, and continue boiling tap water for ice making. Select either a pitcher alternative, POU alternative, or POE to remove the contaminant to safe levels. Disposing of the spent filter cartridges can be challenging. The best treatment option is purified bottled water.

ORGANIC CONTAMINANTS INCLUDING PFAS

Let's talk about the elephant in the room, **per- and polyfluoroalkyl substances (PFAS and perfluorooctanoic acid (PFOA) and perfluorooctanesulfonic acid (PFOS)**.

At the writing of this book, the State of California classified PFOS as a carcinogen in 2021. On March 14, 2023, EPA's science advisory board determined that both PFOA and PFOS are likely to cause kidney and liver cancers. EPA is proposing a National Primary Drinking Water Regulation (NPDWR) to establish Maximum Contaminant Levels (MCLs), for six PFAS in drinking water. EPA is also proposing health-based, non-enforceable Maximum Contaminant Level Goals (MCLGs) of PFOA 4.0 ppt (ng/L) and PFOS 4.0 ppt.

I recommend two American Bar Association articles dealing with PFAS (Cook & O'Reilly, 2021 and Koll & Sheldon, 2023). I also recommend The EPA's rulings.

Again, microbial pathogens have an immediate and life-threatening effect on IC individuals. Inorganic contaminant effects are longer term. The decision to treat your tap water or buy purified bottled water will be determined by absence or presence of PFAS and lead in your tap water and your budget. The technology required to remove PFAS and lead from your tap water is available as POU and POE options. Granular activated carbon, anion ion exchange, and high-pressure membrane technologies have all been demonstrated to remove PFAS, including PFOA, PFOS, PFHxS, HFPO-DA and its ammonium salt, PFNA, and PFBS, from drinking water systems. These treatment technologies can be installed at a water system's treatment plant and are also available through in-home filter options. NSF/ANSI 53 lead certified filters are available as a POU option. A home water softener will remove lead.

REFERENCE LIST

Abubakar, I., et al. "Treatment of cryptosporidiosis in immunocompromised individuals: systematic review and meta-analysis." *British Journal of Clinical Pharmacology*, Hoboken, 2007 Apr;63(4):387–339.

Ali, M.M.M., Mohamed, Z.K., Klena, J.D., Ahmed, S.F., Moussa, T.A.A., and Ghenghesh, K.S. "Molecular characterization of diarrheagenic Escherichia coli from Libya." *The American Journal of Tropical Medicine and Hygiene*, National Institutes of Health, Bethesda, 2012;86:866–871.

Cook, L., and O'Reilly, K. *Regulating PFAS at the Edge of Detection*. American Bar Association, Chicago, June 2021.

https://www.cancer.gov/publications/dictionaries/cancer-terms/def/microorganism

Gaytan-Martinez J., et al, Microbiological findings in febrileneutropena, *Arch Med Res.*, Elsevier, The Netherlands, 2000;31(4):388–392.

Hunter, P.R., and Nichols, G. "Epidemiology and clinical features of Cryptosporidium infection in immunocompromised patients." *Clinical Microbiology Reviews*, Chicago, 2002;15:145–154.

Invik, J., Barkema, H.W., Massolo, A., Neumann, N.F., and Checkley, S. "Total coliform and Escherichia coli contamination in rural well Water: Analysis for passive surveillance." *Journal of Water and Health*, IWA Publishing, London, 2017;15:729–740.

Koll, C., and Sheldon, J. *PFAS Toxicological Risk Status, a Topic of Much Debate*. American Bar Association, Chicago, Sept 1, 2023.

National Research Council Division on Earth and Life Studies; Commission on Life Sciences; Safe Drinking Water Committee. "Radioactivity in drinking water." *Drinking Water and Health: Volume 1*, Chapter VII. National Academies Press (US), Washington, DC, 1977.

Tao, X., et al. "A retrospective study on Escherichia coli bacteremia in immunocompromised patients: Microbiological features, clinical characteristics, and risk factors for shock and death." *Journal of Clinical Laboratory Analysis*, Hoboken, 2020 Aug; 34(8).

Pregnant Mothers and Fetal Health

7

OVERVIEW

You just learned that you're pregnant. Congratulations! So much to think about. In this chapter I take away the need for you to have to "think" about safe drinking water for you and your baby. I provide low-risk, that is, approaching zero-risk, drinking water options for you and your infant during your pregnancy and during the baby's early childhood. I assume that you reside in the United States.

The reason I included pregnant mothers in this book is that the Safe Drinking Water Amendments of 1996 (epa.gov) included children as a sensitive subpopulation with similar, safe drinking water needs as IC individuals. You decide how much you want to know about the Safe Drinking Water Act.

A pediatrician wrote in a scientific journal article entitled, "Is the Water Safe for my Baby?"(Balbus & Lang, 2001), "pediatricians are confronted by questions about drinking water for which their training may not have prepared them adequately." Focus drinking water questions to your pediatrician on contaminants of concern such as fluoride, lead, PFAS, arsenic, and nitrates. Be sure to understand his thinking on Covid-19 and viruses in general. The Covid-19 virus cannot survive water treatment disinfection.

Focus questions on dental fluorosis and safe fluoride levels for you and your fetus during pregnancy and for your newborn until age 8. Your dentist may or may not be familiar with an article published in the *Australian Dental Journal* about fluoride levels in baby formula (Silva & Reynolds, 1996). If your pediatrician or dentist recommends a drinking water treatment technology, always get a second opinion from a "water guy."

DOI: 10.1201/9781003493839-7

INTRODUCTION

Municipal drinking water contaminants that pose a risk to mother and infant are microbial pathogens, chemical contaminants, and radioactivity. In the United States, drinking water is safe to drink except during a pipe break or an unexpected event that triggers a boil water notice. The microbial pathogens *Escherichia coli* (*E. coli*) and *Cryptosporidium parvum* pose the greatest risk, and that risk is generally acceptable for the immunocompetent. Cities with populations of 10,000 or greater are required to perform more frequent testing than cities with less than 10,000, notably for *Cryptosporidium parvum*. More frequent testing is not a solution for the IC and pregnant mothers. The risk to the IC, mother, and baby occurs immediately after an incident, such as a pipe break, and prior to a boil water notice being issued. The chemical contaminants, disinfection by-products, PFAS, arsenic, fluoride, lead, and nitrates pose the greatest risk to mother and baby. Radioactivity in drinking water is geographical. Drinking purified bottled water reduces that risk to near zero.

CONTAMINANTS OF CONCERN

DBPs

Many researchers have studied the effects of DBPs on infants, with no conclusive evidence. Research in 2004 and 2005 cautioned that exposure to elevated levels of drinking water disinfection by-products (DBPs) may cause pregnancy loss. Further studies proved that municipal drinking water in the US was safe for pregnant women. Assume that the MCL of 80 ppb is a safe number. Consult your CCR for violations during Covid and the current levels in your drinking water. Many municipalities over-chlorinated during Covid to ensure 100 percent kill. If your municipal water source is groundwater not under the influence of surface water, the level of DBPs approaches zero and is not a concern. It is only a concern if your municipal's water source is surface water with a high TOC. Use a TOC of greater than 2 as high. The lower the TOC value, the better.

The risk of forming disinfection by-products stems from the fact that a chlorine residual must be maintained at the tap. After treated water leaves the plant, it is delivered via miles of underground pipes subject to leaks and accidental failures, an unacceptable risk to pregnant women and IC individuals. Bottled water poses close to zero risk of DBPs. It is delivered in sanitary sealed containers that were filled under controlled conditions at the bottling plant.

PFAS

The EPA reported in early studies a risk to pregnant mothers from PFAS. The EPA reported adverse health effects observed following oral exposure to such PFAS are growth and development, for example, low birth weight, preterm birth, and gestational age. As a result of early studies on elevated PFAS exposure to infants from breastfeeding, early termination of breastfeeding is recommended, despite the known advantages of breastfeeding (ABA paper).

In addition to cancer, PFAS exposure during pregnancy might be associated with preterm birth complications with the risk of miscarriage and preeclampsia. A number of states have already issued safe drinking water levels for PFAS. In March 2023, the EPA published its proposed PFAS drinking water MCL in the *Federal Register* at 4.0 parts per trillion (ppt). This is the safe level of chronic exposure for everyone, including pregnant mothers. Confirm the PFAS concentration in bottled water before deciding on a brand. Many brands meet a 5 ppt level or better.

Arsenic

Arsenic is carcinogenic as classified by IARC and the National Cancer Institute. Arsenic is rarely detected in municipal drinking water. Private well water should be tested, MCL is 10 ppb. Bottled water value, ND. Bottled water and ND is my strong recommendation.

Cadmium

Cadmium is rarely detected in municipal drinking water. Private well water should be tested. The MCL is 5 ppb. Bottled water is my strong recommendation.

Chromium

Chromium exposure can induce non-life-threatening complications during pregnancy and childbirth (Xia et al., 2016). Chromium is rarely detected in municipal drinking water and ND in purified bottled water. A probable source is private well water. Private well water should be tested. Bottled water ND chromium is my strong recommendation.

Copper

Regulated under the lead-copper rule.

Fluoride

Baby formula contains fluoride which lowers the acceptable fluoride concentration in the drinking water used to reconstitute it. During pregnancy, water quality considerations are for the mother and the fetus. Municipal tap water is considered by many safe for mother and fetus to drink and safe for mixing baby formula because it is pathogen-free. The statement that municipal drinking water is safe for pregnant mothers is mostly correct but does not address infant susceptibility to fluoride and lead. Pathogen-free includes the Covid-19 virus. Viruses are killed by the chlorine disinfection process.

The MCL for fluoride is 4.0 mg/L. The safe level to avoid dental fluorosis in infants is 0.1 mg/L. (Silva & Reynolds, 1996). They arrived at this value because baby formula has fluoride ranging from 0.023 to 3.71 mg/L, averaging 0.240 mg/L. The value of 0.1 mg/L or ppm is no-think concentration value of fluoride in water used to reconstitute baby formula. Call the number on your municipal water's CCR and ask for the value of fluoride in your tap water. You can always have your tap water tested. There are no POU filters available that remove fluoride. Using purified bottled water brands to prepare baby formula is always a better choice. These brands have acceptable fluoride and warrant consideration. Smartwater, Aquafina, Evian, Nestlé Pure Life, Poland Spring, Dasani, ESKA, Sam's Choice, and Icelandic Glacial.

Lead

Speak to your pediatrician about the effects of lead in your drinking water on you and your baby's health and whether a blood lead test is appropriate. The MCL for lead is 15 ppb. The public health goal is zero. The safe level for infants is ND. Most municipalities meet an ND level for lead. It's your private well and/or house plumbing that might be of concern. Purified bottled water is ND for lead and does not flow through your house's plumbing.

Nitrates

The risk of nitrates is to infants below age 6. The most severe consequence is infant methemoglobinemia (blue baby syndrome), when there is not enough oxygen in the blood caused by high levels of nitrates in baby's formula. Nitrates above its 10 mg/L -MCL is rare in US drinking water. The greatest risk is from a contaminated well. Nitrogen fertilizers are generally used to enrich soils since nitrates are a critical source of nitrogen for plants. Rain and irrigation can transport the nitrates through the soil to the groundwater. Septic tank effluent is another source when a septic tank is located too close to a well.

Pesticides

The presence of pesticide residues in drinking water can be a major risk to maternal and fetal well-being. Pesticides are ND in most municipal drinking water. Pesticides are ND in purified bottled water. If your water comes from a private well, pesticides may be present. Your well water should be tested.

Radiation

Everyone is exposed to some natural radiation. Radioactivity, ionizing radiation, is present to some degree in all drinking water. Its concentration and composition depend on the radiochemical composition of the geology through which the raw water may have passed that comes from both cosmic rays and terrestrial sources. The average background dose in the United States is between 100 and 200 mrem/yr. A small proportion of this unavoidable background radiation comes from drinking water that contains radionuclides.

Radiation exposure is a risk in some states, but the greater risk to mother and infant is from medical imaging, that is, plain radiograph (x-ray) and computed tomography (CT). Check your CCR for radioactivity. Your CCR indicates the presence of uranium or radon in your drinking water. Take your CCR's radioactivity level with you, and show it to the technician so she/he is aware of your exposure.

The developing fetus is exposed to radiation from radionuclides in drinking water for nine months. Thus, the total dose accumulated by the fetus will be very small. Furthermore, although the fetus is sensitive to the effects of radiation in some stages of development, these periods are sharply limited and extremely short. For this reason, too, the total dose administered that could possibly have developmental and teratogenic effects would be extremely small. Current concentrations of radionuclides in drinking water led to doses of about one five-thousandth of the lowest dose at which a developmental effect has been found in animals. Therefore, the developmental and teratogenic effects of radionuclides would not be measurable (NRC, 1977).

MICROBIAL CONTAMINANTS

The risk to you and your infant from pathogens in municipal drinking water is extremely low but not zero. There is zero risk in purified bottled water. The highest risk is from a private well. You are more likely to get diarrhea from preparing food. If a drinking water microbial contamination event occurs, your municipality will issue a boil water notice. You then can either boil your water as directed or drink bottled water. The risk, however, to you and your baby occurs immediately after the event, a pipe break, for example, and before the boil water notification is issued. A larger water provider, 10,000 people or greater than one million gallons per day (mgd), is less likely to experience an event.

See microbial contaminants in Chapter 3. *Escherichia coli* (*E. coli*) infection, in rare cases, can cause miscarriage or preterm delivery (Ovalle & Levancini, 2001).

MITIGATION STRATEGIES

The decision to treat your tap water or buy purified bottled water will be determined by the absence or presence of PFAS and lead in your tap water and your budget. Activated carbon, anion exchange (AIX), and high-pressure membrane technologies have all been demonstrated by the EPA to remove PFAS, including PFOA, PFOS, PFHxS, HFPO-DA and its ammonium salt, PFNA, and PFBS, from drinking water systems. These treatment technologies can be installed at a water system's treatment plant and are also available through in-home filter options. Lead is removed by your softener. No softer? Consider an NSF certified lead filter.

One way to assess your risk of boil water notices is to review your utilities' CCR for the last five years to determine the frequency of boil water events. No events? Great. Many events? Consider an NSF-certified absolute POU bacteria filter for the faucet that supplies drinking water for you and your baby. Do this even if you are breastfeeding to minimize any bacterial infections during a boil water notice.

A strategy for pregnant mothers who are not breastfeeding is to use a purified bottled water with a PFAS concentration of 5 ppt both as your drinking water source and to mix baby formula until the quality of tap water is confirmed for the contaminants of concern. Many bottled waters company's post a water quality report and/or a water analysis online to help with your decision. You can search bottled waters brands online and choose the one that is right for your baby, your budget, and you.

The next step is to check out your water utility's CCR for the contaminants of concern. You can always call the utility and speak with a person if you're not sure that the CCR answers your questions. Another important point is that the water quality stated in your CCR is the quality of water leaving the water treatment plant and not the quality of the water coming out of your faucet. You may want to have your water tested for the contaminants of concern to confirm any changes in water quality caused by the distribution system.

Decision time. If you're breastfeeding, most of the drinking water choices go away except for lead. Since lead can be transferred from mother to baby, the amount of lead in the water that you drink should be ND. Boiling water does not remove lead. Drink and cook only with cold water. Warm or hot water may have higher lead. But first confirm that you do not have a lead service line. If no, great. If yes, replace it. Call your water authority to evaluate your options. If you can afford it, just replace the line.

Another option is POU water treatment on the faucet or pitcher that you use for drinking water. Choose a treatment that has NSF/ANSI Standard 53 certification for lead removal and 42 for particulate removal. This is important. Testing a treatment technology to NSF/ANSI standards is not the same as NSF/ANSI certification.

REFERENCE LIST

Balbus, J.M., and Lang, M.E. "Is the water safe for my baby." *Pediatric Clinics of North America*, New York, 2001 Oct;48(5):1129–1152.

National Research Council Division on Earth and Life Studies; Commission on Life Sciences; Safe Drinking Water Committee. "Radioactivity in drinking water." In *Drinking Water and Health: Volume 1,* Chapter VII. National Academies Press, Washington, DC, (US), 1977.

Ovalle, A., and Levancini, M. "Urinary tract infections in pregnancy." *Current Opinion in Urology*, NIH, Baltimore, MD, 2001 Feb;11(1):55–59.

Silva, M., and Reynolds, E.C. "Fluoride content of infant formulae in Australia." *Australian Dental Journal*, Hoboken. NJ, 1996;41(1):37–42.

Tao, X., et al. "A retrospective study on Escherichia coli bacteremia in immune compromised patients: Microbiological features, clinical characteristics, and risk factors for shock and death, *Journal of Clinical Laboratory Analysis*, Hoboken, 2020; 34(8).

Xia Wei, et al. "A case-control study of maternal exposure to chromium and infant low birth weight in China." *Chemosphere*, St louis, MO, 2016 Feb;1484–1489.

Final Thoughts

8

SUMMARY

Safe Drinking Water for the Immunocompromised is the first book written for the millions of immunocompromised individuals about safe drinking water choices. It delineates drinking water choices as municipal tap water, bottled water, and, for 15 million people in the United States, a private groundwater well. The greatest risk to IC individuals is microbial, pathogenic *Escherichia coli*, which is a pathogen in drinking water linked to a variety of diseases. *Escherichia coli* O157:H7 can cause kidney failure and death in immunocompromised individuals. *Cryptosporidium parvum* is associated with severe life-threatening illness among immunocompromised individuals. Legionnaires' disease is not usually transmitted through drinking water. Viral infections transmitted through drinking water are extremely rare because of US disinfection practices. Pathogenic fungi are opportunistic and can cause fungal infection disease in patients with immunocompromised conditions. Inorganic compounds of concern to IC individuals are heavy metals including arsenic, cadmium, chromium, fluoride, mercury nitrates, and radioactive compounds. The organic compounds of most concern are PFAS, pesticides, and disinfection by-products. Mitigation strategies for microbial include boiling tap water for drinking, ice making, and teeth brushing and/or absolute filtration. The best option from a water quality perspective is purified bottled water. Mitigation strategies for inorganic contaminants include POU/POE nano-filtration or reverse osmosis with the most desirable, purified bottled water. Mitigation options for PFAS are POE/POU granular activated carbon, anion exchange, or high-pressure reverse osmosis with, again, the most effective, purified bottled water.

NEXT STEPS

One of the reasons I wrote this book was to bring attention that the immuno-compromised community has been forgotten by the EPA. The time is now for the EPA to look at the best way to serve the needs of IC individuals. It's time for the Bottled Water Association to consider offering an affordable line of bottled water for IC individuals and pregnant women. The opportunity is there for a bottled water provider to offer a special brand of bottled water to the IC community.

The time is now for schools of medicine to offer a curriculum addressing the water quality needs of IC individuals.

Who will step up?

Glossary of Terms

absolute filtration: The pore opening size of the filter. Filter media with an exact and consistent pore have an absolute rating.

acidity: Neutralizing power of a water. Adding acid to water lowers its pH.

alkalinity: A measure of the acid-neutralizing power of water, its capacity to absorb acidity without significant pH change.

aluminum: A naturally occurring element that is used in water treatment chemicals.

anion: A negatively charged ion.

arsenic: A naturally occurring element linked to cancer of the bladder, lungs, skin, kidney, liver, and prostate.

atrazine: A herbicide that causes cancer and pregnancy issues. It must be avoided by IC and pregnant mothers.

barium: An alkaline earth metal linked to blood pressure increase.

base: Another name for an alkaline solution with a pH greater than 7. Sodium hydroxide is a strong base.

boil water order: Drinking water providers must inform consumers no later than 24 hours after an event that their water is contaminated, usually by microorganisms. The water should be boiled before drinking. *IC individuals cannot risk the 24-hour lag time.*

bottled artesian water (artesian well water): Bottled water that originates from a well drilled into a confined aquifer.

bottled groundwater: Water from a protected underground water source.

bottled mineral water: Water that originates from a protected underground water source with a TDS concentration not less than 250 mg/L. Minerals and trace elements must come from the water source and not added later.

bottled purified water: Water that meets EPA's primary drinking water standards, further treated by distillation, deionization, or reverse osmosis.

bottled spring water: Water that originates from an underground spring.

bottled sterile water: Bottled water that meets sterility testing requirements.

bottled tap water: Bottled water that uses tap water as its source. The water may be either bottled as is or further treated.

calcium: Together with magnesium, a major component of hardness.

cation: A positively charged ion.

chloride: An anion that gives water a salty taste. *High chloride content water may not be suitable for certain IC individuals.*

chlorine: A chemical added after water treatment, usually in the form of hypochlorite (bleach), to kill microbial contaminants and pathogens.

coliform/coliform group of bacteria: A group of related bacteria usually found in the intestines of warm-blooded animals. Used to indicate fecal contamination.

concentration: A measure of the amount of a substance dissolved in drinking water.

conductivity: A measure of the ability of an aqueous solution (water solution) to conduct an electric charge.

***Cryptosporidium parvum*:** Also "*Crypto*." An intestinal protozoan parasite that lives in the intestine of humans and animals that causes severe self-limited diarrhea and pain in healthy (immunocompetent) persons and severe, life-threatening disease in immunocompromised individuals which could last a lifetime.

Cryptosporidiosis: A disease with no cure caused by the parasite *Cryptosporidium parvum*. Municipal water providers for cities with a population greater than one million gallons per day (1 MGD) are required to test for *cryptosporidium parvum*.

dental fluorosis: A condition caused by contact of children under the age of 8 with too much fluoride, resulting in changes in the appearance of tooth enamel, usually white or brown speckles on the teeth.

deionized bottled water: Bottled water further treated by the ion exchange process.

disinfection: A process whereby chemicals are used to kill pathogens.

Disinfection by-products: Chemical by-products formed when water is disinfected with chlorine. The compounds formed are trihalomethanes (THMs) and five haloacetic acids (HAA5). *DBPs are a concern for IC individuals.*

distillation: A water purification process during which water is evaporated and its vapor condensed and collected as pure water.

distilled bottled water: Bottled water that has been purified by distillation. *Distilled bottled water is a preferred choice for IC individuals.*

***Escherichia coli (E.Coli)*:** A bacterium that lives in the intestines of warm-blooded animals. *E coli O157:H7* can cause bloody diarrhea.

fecal coliform bacteria: Found in the intestines of warm-blooded animals.

fluoride: An anion sometimes added to drinking water to prevent tooth decay. Can cause dental fluorosis in infants.

***Giardia lamblia*:** A parasite that causes the diarrheal disease giardiasis.

Giardiasis: A protozoan infection of the small intestines. *Giardiasis is of concern to IC individuals.*

granular activated carbon: Used to remove chlorine and reduce organics in drinking water pitcher, POU, and POE systems.

haloacetic acids (HAAs): Disinfection by-products formed during the chlorination of water containing natural organic matter.

hardness: A characteristic of water caused by high concentrations of calcium and magnesium salts. Hard water makes lathering with soap difficult and causes scale to form on cooking pots. It causes discoloration around bathtub drains.

heavy metals: Cadmium, chromium, lead, mercury, nickel, and zinc.

immunocompromised: Individuals with immune systems that lack the ability to fight off infections and disease.

immunocompetent: Individuals with a normal functioning immune system.

inorganic matter: Matter originating from mineral sources.

ion: An atom with more or fewer electrons than protons.

iron: A naturally occurring element found in many rocks and soils. Dissolved iron is found in most water supplies. Concentrations greater than 0.3 mg/L cause red stains.

lead: A regulated metallic element that is banned in pipes. It may cause a range of health effects in children including mental development.

***Legionella pneumonphila*:** A bacterium of the family *Legionella* that causes Legionnaires' disease, a serious type of pneumonia.

Legionnaires' disease: A pneumonia caused by the bacterium *Legionella pneumonphila*. Most healthy individuals do not contract Legionnaires' disease. Outbreaks of Legionnaires' disease is usually associated with large complex water systems found in hospitals, hotels, and cruise ships.

magnesium: A cation together with calcium that causes hard water.

manganese: A cation that causes undesirable taste and a black discoloration of water and usually clothes.

maximum contaminant level (MCL): The level of a contaminant that is allowed in drinking water. MCLs are enforceable standards.

maximum contaminant level goal (MCLG): The maximum level of a contaminant in drinking water at which no known or anticipated adverse effect on the health of persons would occur. MCLGs are non-enforceable public health goals.

mercury: A highly toxic heavy metal.

National Sanitation Foundation/American National Standards Institute (NSF/ANSI): A third-party entity that develops standards for clean water products. Products are rigorously tested in NSF laboratories to prove that they do what they claim.

natural organic matter (NOM): Organic matter found in surface waters caused by decaying vegetation, known as "humic" materials. NOM reacts with chlorine to form disinfection by-products that may be of concern to IC individuals.

nitrate and nitrite: Nitrate is often used in fertilizer, and it also is a by-product of animal and human waste. These ions can cause methemoglobinemia, which can be fatal to infants.

nominal filtration: A filter rating that is generally discouraged for IC individuals. Only filtration products with an absolute rating should be used.

non detectable (ND): An analytical sample where the concentration is deemed lower than what could be detected using the analytical method employed.

opportunistic: in medicine, a microorganism hat affects patients when the immune system is depressed.

opportunistic infection: an infection caused by a microorganism that rarely affects patient except when the immune system is depressed.

opportunistic pathogen: non pathogenic bacteria dormant until host immune system is depressed.

organic compound: A carbon containing compound derived from living organisms.

ozonation: The process of applying ozone as a disinfectant. Ozonation is used extensively in Europe and in bottled water treatment.

palatable water: Water that tastes good.

Perfluorooctanoic acid (PFOA) and Perfluorooctanesulfonic acid (PFOS): Part of a large group of man-made chemicals used in the production of industrial and chemical products until around the year 2000. Likely to cause cancer. Other potential health effects include developmental effects to fetuses during pregnancy, and immune effects. Can be present in public drinking water systems and private drinking water wells. The EPA proposed an MCLG of 5 ppt. *All IC individuals should consider no-risk water treatment options for PFASs.*

pH: A measure of the acidity, neutrality, or alkalinity of a water, where 7 is neutral.

point of entry (POE): Point of entry water treatment means treating all water entering a house or building.

point of use (POU): Under-the-sink water treatment.

potable water: Water that may be safe for IC individuals to drink.

primary information sources: Records of events or evidence as they are first described or happened without any interpretation or commentary.

purified bottled drinking water: Source water that meets EPA primary drinking water standards is treated by distillation, deionization, or reverse osmosis. Bottled water treated only by deionization may not be safe for IC individuals. Only water treated by distillation or reverse osmosis is absolutely safe under all circumstances.

reverse osmosis (RO) water treatment systems: A pressure-driven membrane separation technology that removes ions, salts, dissolved solids, and most organics found in drinking water. Can be used as either a POE or POU treatment option.

radionuclide: A material that produces radiation and an increased risk of cancer.

radon: A gaseous radioactive element created from the radioactive decay of radium. Radon is a known carcinogen.

roentgen equivalent man: The roentgen equivalent man (rem) is a CGS unit of equivalent dose, effective dose, and committed dose, which are dose measures used to estimate potential health effects of low levels of ionizing radiation on the human body.

salt: An ionic compound consisting of a cation and anion. Sodium chloride is an example of a salt.

sodium: A metallic element always combined in a compound. Of concern to individuals with circulatory and cardiac conditions and high blood pressure in pregnant women.

sulfate: May act as a laxative in drinking water. Maximum concentration is 250 mg/L.

total dissolved solids: A measure of the dissolved inorganic matter in water.

total coliforms: The group of bacteria used as indicator organisms for fecal contamination.

total organic carbon: A measure of the organic matter in drinking water that can react with chlorine during disinfection to form disinfection by-products.

trihalomethane (THM): Any of several derivatives of methane formed during the disinfection of water by chlorine that exhibit carcinogenic potential. Of particular concern to cancer patients and all IC individuals.

turbidity: A measure of a water's cloudiness or clarity. Not expressed as mg/L or ppm but as Nephelometric Turbidity Units (NTUs).

virus: A strand of RNA or DNA.

zinc: An essential trace nutrient. May give water a metallic taste.

Index

For Product Safety Concerns and Information please contact our EU
representative GPSR@taylorandfrancis.com
Taylor & Francis Verlag GmbH, Kaufingerstraße 24, 80331 München, Germany

www.ingramcontent.com/pod-product-compliance
Ingram Content Group UK Ltd.
Pitfield, Milton Keynes, MK11 3LW, UK
UKHW021112180425
457613UK00005B/58